Remarks on some Fossil Impressions

1854

JOHN COLLINS WARREN

TABLE OF CONTENTS

FOREWARD

The principal part of these remarks were made at the meetings of the Boston Society of Natural History. A portion of them also have been printed in the Proceedings of the Society.

The object of this publication is to afford to those who are not members of the Society an opportunity of obtaining some knowledge of Fossil Impressions, which they might not be able to obtain elsewhere so conveniently.

Some account of the Epyornis seems to be very properly connected with Ornithichnites.

The first of these papers was written in October, 1853; the others in the earlier part of the present year.

THE EPYORNIS

In the course of the year 1851, an account was circulated of the discovery of an immense egg, or eggs, in the Island of Madagascar. The size of the eggs spoken of was so disproportionate to that of any previously known, that most persons received the account with incredulity; and, I must confess, I was one of this number. Being in Paris soon after hearing of this report, I made inquiry on the subject, and was surprised to learn, that the great egg was actually existing in the Museum of Natural History in Paris. In a few days I had an opportunity of seeing a cast of it in the hands of the artist, M. Strahl, of whom I solicited one. He informed me that it could not be obtained at that moment; but that, if my request were made known to the Administration of the Museum, he had no doubt they would accede to it. I accordingly did apply, and also presented them with the cast of a perfect head of Mastodon Giganteus; and they very liberally granted my request.

The distinguished naturalist, Professor Geoffroy St. Hilaire, the second of that honorable name, has made a statement to the Academy of Sciences, which, though only initiatory, contains many facts of a very interesting nature, some of which I have had an opportunity of verifying; and to him we are indebted for a greater part of the others.

The eggs sent to me are, in number, two; one of which was purchased by M. Abadie, captain of a French vessel, from the natives. Another was soon afterwards found, equal in size. A third egg was discovered in an alluvial stratum near a stream of water, together with other valuable relics of the animal which had probably produced them; but, unfortunately, it was broken during transportation. Of the two eggs, one is of an ovoid form, having much the shape of a hen's egg; and the other is an ellipsoid.

The ovoid egg is of enormous size, even when compared with the largest egg we are acquainted with. Its long diameter exceeds thirteen inches of our English measure, its short diameter eight, and its long circumference thirty-three inches. Its capacity is thought to be equal to eighteen liquid pints, or to be six times greater than that of the largest egg known to us (the ostrich), although but twice its length. It is said to be equal to a hundred and forty-eight hen eggs. The ellipsoid egg has its longest diameter somewhat less than that of the ovoid; its short diameter nearly equals that of the other egg, being more than eight inches. The third egg, although broken, has been very useful to science, by displaying the thickness of the shell, which is about one-tenth of an inch.

The bones, of which I have received the casts, are three in number, and of great interest. One of them is a characteristic fragment of the upper part of a fibula; the other two, still more interesting, as enabling us to determine the class and genus of the animal to which they belong, exhibit the extremities of the right and left tarso-metatarsal bones. The former is somewhat broken; the latter is nearly perfect, and exhibits the triple division of the inferior extremity of the bone into the three trochleæ or pulley-shaped processes of the struthious birds. It might be mistaken for a bone of the great Dinornis, but is distinguished from this by the flatness of the portion above the trochleæ. Still less is it one of the bones of the ostrich, its three pulleys being separated from each other by distinct intervals; whereas the pulleys of the ostrich have only one such separation, constituting two distinct eminences.

M. Geoffroy St. Hilaire considered himself justified, from these and other facts, in deciding this bone to belong to a bird of a new genus, to which he gives the name of Epyornis, from αἰπύς, high, tall, and ὄρνις, bird; and, as probably it is a specimen of the largest animal of the family, he affixes the specific name of maximus.

The size of this bird, inferred from that of its egg, would be vastly superior to that of the ostrich. But if we notice the comparative size of the trochleated extremity of the tarso-metatarsal bone, we shall see that its height would be greatly exaggerated by adopting such a basis for its establishment; in fact, it would not probably exceed a height double that of the ostrich. And, though it must have been superior to that of the Dinornis maximus of Prof. Owen, it might perhaps excel it only by the difference of two or three feet. A bird of twelve or thirteen feet in height would, however, if we stood in its presence, appear enormous, and must have greatly astonished and terrified the natives of Madagascar. Whether it now exists is uncertain, as it may possibly have a habitation in the wild recesses of the island, which have never yet been visited by any European traveller.

The credit of most of the observations and discoveries relating to this remarkable bird is attributable to French naturalists;[A] and it seems to be a duty devolving on English and American navigators to complete the history thus happily begun, and to tell us whether the Epyornis still exists in the mountain-forests of Madagascar, or at least present us with its extraordinary relics.

FOSSIL IMPRESSIONS.I

Ichnology, a newly created branch of science, takes its name from the Greek word ἴχνος , a track or footstep, and the tracks themselves have been denominated Ichnites, or, when they refer to birds only, Ornithichnites, from ὄρνις, a bird. And this last term has by custom been generally applied to ancient impressions, though not correctly.

Geology has revealed to us not only the remains of animals and vegetables, but the impressions made by them during their lives, and even the impressions of unorganized bodies. The first notice of these appearances was, as often happens, regarded with indifference or scepticism; but their number and variety enlightened the public mind, and opened a new source of information and improvement.

The first remarkable observation made on fossil footsteps was that of the Rev. Dr. Duncan, of Scotland, in 1828. He noticed, in a new red sandstone quarry in Dumfriesshire, impressions of the feet of small animals of the tortoise kind, having four feet, and five toes on each foot. They were seen in various layers through a thickness of forty feet or more.

Sandstone, in which these impressions are principally discovered, is a rock composed chiefly of siliceous and micaceous particles cemented together by calcareous or argillaceous paste, containing salt, and colored with various shades of the oxide of iron, particularly the red, gray, brown. It has been remarked by Prof. H. D. Rogers, that the perfection of the surface containing fossil footmarks is often attributable to a micaceous deposit. The layers of sandstone have been formed by deposits from sea-water, dried in succession; such layers are also seen in the roofing slate. These deposits on the shores of the ocean, having in a soft condition received the impressions of the feet of birds, other animals, vegetables, and also of rain-drops, under favorable circumstances dried, hardened, and formed a rock of greater or less solidity. Our colleague, Dr. Gould, has exhibited to us a specimen of dried clay from the shores of the Bay of Fundy, containing beautiful impressions, recently made, of the footsteps of birds. The particles brought by the waves, and deposited in the manner described, were derived from the destruction of other rocks previously existing, particularly granite and flint, or silex, the shining atoms of which compose no small part of the sandstone rock.

It is easy to conceive, that, while these deposits were taking place in the soft condition, portions of vegetable matters might become intermixed; and that these, with the impressions of the feet and other parts of animals and unorganized substances, might be preserved by the process of desiccation. The agency of internal heat may have also been employed in some cases in baking and hardening these crusty layers.

The sandstone rock, though in some places actually in a state of formation at the present time, lies in such a manner in the earth's crust as to indicate an immense antiquity. The age of these beds varies in different situations. The sandstone rocks which contain the greater part of the impressions are called new red sandstone, to distinguish them from the old red, which is of a greater age. The deposits on Connecticut River may not be attributed to the action of this river, but are of higher antiquity, probably, than the river itself, and proceeded from the waves of an ancient sea, existing in a state of the surface of the globe very different from that of the present day.

In 1834, tracks were discovered near Hildberghausen in Saxony, to which Prof. Kaup, of Darmstadt, gave the name of Chirotherium, from the resemblance to the impressions of the human hand. On a subsequent examination, Prof. Owen preferred the name of Labyrinthodon, from the resemblance of the folds in the teeth to the convolutions of the brain.

Various other instances of impressions were seen; and, in the year 1835, Dr. Deane and Mr. Marsh, residents of Greenfield, noticed impressions resembling the feet of birds in sandstone rocks of that neighborhood. These observations having come to the knowledge of President Hitchcock, of Amherst College, that gentleman began a thorough investigation of the subject, followed it up with unremitted ardor, and has, since 1836 (the date of his first publication), laid before the public a great amount of ichnological information, and really created a new science. Dr. Deane, on his part, has not been idle: besides making valuable discoveries, he has written a number of excellent papers to record some portion of his numerous observations.

In 1837, at the request of my friend Dr. Boott, I carried to London, for the Museum of the Royal College of Surgeons, various scientific objects peculiar to this country; among which were a number of casts of Ornithichnites.

These casts were kindly furnished me by President Hitchcock, and the Government of the Royal College thereon voted to present to President Hitchcock and Amherst College casts of the skeleton of the famous Megatherium of South America. These casts were packed, and sent to be embarked in a ship destined for Boston, but were unluckily delivered to a wrong shipping house in London, and I lost sight of them for some time. They were at length discovered. After remaining in this situation for more than a year, they were sold at public auction; and, notwithstanding many efforts on my part, I was unable to obtain and transmit them to Amherst College.

The fossil impressions which have been distinguished in various places in the new red sandstone are those of birds, frogs, turtles, lizards, fishes, mollusca, crustacea, worms, and zoophytes. Besides these, the impressions made by rain-drops, ripple-marks in the sand, coprolites or indurated remains of fæces of animals, and even impressions of vegetables, have been preserved and transmitted from a remote antiquity. No authentic human impressions have yet been established; and none of the mammalia, except the marsupials.(?) We must, however, remember that, although the early paleontology contains no record of birds, the ancient existence of these animals is now fully ascertained. Remains of birds were discovered in the Paris gypsum by Cuvier previous to 1830. Since that time, they have been found in the Lower Eocene in England, and the Swiss Alps; and there is reason to believe that osseous relics may be met with in the same deposits which contain the foot-marks. Most of the bird-tracks which have been observed, belong to the wading birds, or Grallæ.

The number of toes in existing birds varies from two to five. In the fossil bird-tracks, the most frequent number is three, called tridactylous; but there are instances also of four or tetradactylous, and two or didactylous. The number of articulations corresponds in ornithichnites with living birds: when there are four toes, the inner or hind toe has two articulations, the second toe three, the third toe four, the outer toe five. The impressions of the articulations are sometimes very distinct, and even that of the skin covering them.

President Hitchcock has distinguished more than thirty species of birds, four of lizards, three of tortoises, and six of batrachians.

The great difference in the characters of many fossil animals from those of existing genera and species, in the opinion of Prof. Agassiz, makes it probable that in various instances the traces of supposed birds may be in fact traces of other animals, as, for example, those of the lizard or frog. And he supports this opinion, among other reasons, by the disappearance of the heel in a great number of Ornithichnites.

D'Orbigny, to whom we are indebted for the most ample and systematic work on Paleontology ("Cours Elémentaire de Paléontologie et de

Géologie," 5 vols. 1849-52), does not accept the arrangement of President Hitchcock. He objects to the term Ornithichnites, and proposes what he considers a more comprehensive arrangement into organic, physiological, and physical impressions. Organic impressions are those which have been produced by the remains of organized substances, such as vegetable impressions from calamites, &c. Physiological impressions are those produced by the feet and other parts of animals. Physical impressions are those from rain-drops and ripple-marks; and to these may be added coprolites in substance. This plan of D'Orbigny seems to exclude the curious and interesting distinctions of groups, genera, and species; in this way diminishing the importance of the science of Ichnology.

Fossil impressions have been found on this continent in the carboniferous strata of Nova Scotia, and of the Alleghenies; in the sandstone of New Jersey, and in that of the Connecticut Valley in a great number of places, from the town of Gill in Massachusetts to Middletown in Connecticut, a distance of about eighty miles.

A slab from Turner's Falls, obtained for me by Dr. Deane in 1845, measuring two feet by two and a half, and two inches in thickness, contains at least ten different sets of impressions, varying from five inches in length to two and a half, with a proportionate length of stride from thirteen inches to six. All these are tridactylous, and represent at least four different species. In most of them the distinction of articulation is quite clear. The articulations of each toe can readily be counted, and they are found to agree with the general statement made above as to number. The impressions are singularly varied as to depth; some of them, perfectly distinct, are superficial, like those made by the fingers laid lightly on a mass of dough, while others are of sufficient depth nearly to bury the toes; some of the tracks cross each other, and, being of different sizes, belong to animals of different ages or different species. There is one curious instance of the tracks of a large and heavy bird, in which, from the softness of the mud, the bird slipped in a lateral direction, and then gained a firm footing; the mark of the first step, though deep, is ill-defined and uncertain; the space intervening between the tracks is superficially furrowed; in the settled step, which is the deepest, the toes are very strongly indicated. On the same surface are impressions of nails, which may have belonged to birds or chelonians.

The inferior surface of the same slab exhibits appearances more superficial, less numerous, but generally regular. There are three sets of tracks entirely distinct from each other; two of them containing three tracks, and one containing two,—the latter being much the largest in size. In addition, there is one set of tracks, which are probably those of a tortoise. These marks present two other points quite observable and interesting. One is that they are displayed in relief, while those on the upper surface are in depression. The relief in this lower surface would be the cast of a cavity in the layer below; so the depressions in the upper surface would be moulds of casts above. The second point is the non-correspondence of the upper and lower surfaces; i.e. the depressions in the upper surface have not a general correspondence with the elevations on its inferior surface. The tracks above were made by different individuals and different species from those below. This leads to another interesting consideration, that in the thickness of this slab there must be a number of different layers, and in each of them there may be a different series of tracks.

To these last remarks there is one exception: the deep impression in which the bird slipped in a lateral direction corresponds with an elevation on the lower surface, in which the impression of these toes is very distinctly displayed, and even the articulations. Moreover, one of the tracks on the inferior surface interferes with the outer track in the superior, and tends in an opposite direction, so that this last-described footstep must have been made before the other. It is also observable, that, while all the other tracks are superficial, this last penetrates the whole thickness of the slab; thus showing that the different deposits continued some time in a soft state.

On the surfaces of this slab, particularly on the upper, there are various marks besides those of the feet, some of which seem to have been made by straws, or portions of grass, or sticks; and there is a curved line some inches in length, which seems to have arisen from shrinkage.

In the collection of Mr. Marsh,[B] there were two slabs of great size, each measuring ten by six feet, having a great number of impressions of feet, and about the same thickness as the slab under examination. One of these presented depressions; and the other, corresponding reliefs. These very interesting relations were necessarily parted in the sale of Mr. Marsh's collection; one of them being obtained for the Boston Society of Natural History, and the other for the collection of Amherst College.

The Physical Impressions, according to Professor D'Orbigny, are of three kinds, viz.: 1st, Rain-drops; 2d, Ripple-marks; and 3d, Coprolites. I have a slab which exhibits two leptodactylous tracks very distinct, about an inch and a half long, surrounded by impressions of rain-drops and ripple-marks. Another specimen exhibits the impressions of rain in a more distinct and remarkable manner. The imprints are of various sizes, from those which might be made by a common pea to others four times its diameter; some are deep, others superficial and almost imperceptible. They are generally circular, but some are ovoid. Some have the edge equally raised around, as if struck by a perpendicular drop; and others have the edge on one part faintly developed, while another part is very sharp and well defined, as if the drop had struck obliquely. It has been suggested, that these fossil rain-drops may have been made by particles of hail; but I think the variety of size and depth of depression would have been more considerable if thus made.

Although we have necessarily treated the subject of fossil footmarks in a very brief way, sufficient has been said to show that this new branch of Paleontology may lead to interesting results. The fact that they are, in some manner, peculiar to this region, seems to call upon our Society to obtain a sufficient number of specimens to exhibit to scientific men a fair representation of the condition of Ichnology in this quarter of our country; and we have therefore great reason to congratulate ourselves, that, through the vigilance and spirit of our members, the Society has the expectation of obtaining a rich collection of ichnological specimens.

FOSSIL IMPRESSIONS.II

Since writing the preceding article, I have been able to obtain, through the kindness of President Hitchcock, a number of additional specimens of fossil impressions. By the aid of these, I may hope to give an idea of the system of impressions, so far as it has been discovered, without, however, attempting to enter into minute details. For these, I would refer to the account of the "Geology of Massachusetts," by President Hitchcock; to his valuable article published in the "Memoirs of the American Academy;" and to his geological works generally.

The numerous tracks which have been assembled together in the neighborhood of Connecticut River have afforded an opportunity of prosecuting these studies to an extent unusual in the primitive rocky soil of New England. These appearances are not, indeed, wholly new. Such traces had been previously met with in other countries; but, in their number and variety, the valley of the Connecticut abounds above all places hitherto investigated.

Twenty years have elapsed since the study of Ichnology has been prosecuted in this country; and, in this period of time, about forty-nine species of animal tracks have been distinguished in the locality mentioned, according to President Hitchcock; which have been regularly arranged by him in groups, genera, and species.

I propose now to lay the specimens, recently obtained, before the Society, as a slight preparation for the more numerous and more valuable articles which they are soon to receive.

The traces found on ancient rocks, as has been shown in the previous article, are those of animals, vegetables, and unorganized substances. The traces of animals are produced by quadrupeds, birds, lizards, turtles, frogs, mollusca, worms, crustacea, and zoophytes. These impressions are of various forms: some of them simple excavations; some lines, either straight or curved, and others complicated into various figures.

President Hitchcock has based his distinctions of fossil animal impressions on the following characters, viz.:—

1.Toes thick, pachydactylous; or thin, leptodactylous.
2.Feet winged.
3.Number of toes from two to five, inclusive.
4.Absolute and relative length of the toes.
5.Divarication of the lateral toes.
6.Angle made by the inner and middle, outer and middle toes.
7.Projection of the middle beyond the lateral toes.
8.Distance between tips of lateral toes.
9.Distance between tips of middle and inner and outer toes.
10.Position and direction of hind toe.
11.Character of claw.
12.Width of toes.
13.Number and length of phalangeal expansions.
14.Character of the heel.
15.Irregularities of under side of foot.
16.Versed sine of curvature of toes.
17.Angle of axis of foot with line of direction.
18.Distance of posterior part of the foot from line of direction.
19.Length of step.
20.Size of foot.
21.Character of the integuments of the foot.
22.Coprolites.
23.Means of distinguishing bipedal from quadrupedal tracks.

By these characters, President Hitchcock has distinguished physiological tracks, or those made by animated beings, into ten groups provisionally. To these may be added, "organic impressions," made by organized bodies; and the impressions made by inanimate bodies, called "physical impressions."

The specimens under our hands enable us to give some notion of the distinctions which characterize the greater part of these groups.

GROUP FIRST STRUTHIONES

The ostrich-tracks present a numerous natural and most remarkable group; remarkable from the great size of some species,—all of them tridactylous and pachydactylous. The ostrich of the Old World has only two toes, but this family exists in South America at the present time under the name of Rhea Americana; and tracks of an animal, probably of the same family, are found in the numerous impressions near Connecticut River,—all of them having three toes in front, and the rudiment of a fourth behind.

This group contains a number of genera. The First Genus, denominated Brontozoum, presents the tracks of a most extraordinary bird. These tracks appear less questionable since the discovery in Madagascar of the eggs of the Epyornis.

The tracks of the largest species, the Brontozoum Giganteum, are four times the magnitude of those made by the existing ostrich of Africa. They are very numerous, and congregated together. The foot of the Brontozoum Giganteum, including the inferior extremity of the tarso-metatarsal bone, which makes a part of the foot, measures in our specimen twenty inches; in the Mastodon Giganteus, the foot measures twenty-seven inches; the width also is less, being ten inches across the metacarpals, while that of the Mastodon is twenty-two: but the one is a bird, the other a quadruped. The toes are three in number, and present the same divisions with existing birds; the inner toe having three, the middle four, the outer five phalanges. Some of the articulations of the toes of this noble specimen are remarkable for the manner in which they illustrate the mode of formation of the tracks. These phalanges have become separated from the solid rock in which they were encased, so as to be removable at pleasure; and they thus show that the whole foot is not a simple impression in the rock which contains it, but a depression filled by foreign materials, i.e. by sand, clay, and other relics of pre-existing rocks. These materials had been gradually deposited in the mould formed by the bird's foot, and are therefore independent of this rock, in the same way as the plaster-of-Paris cast of a tooth, or any other body, is independent of the mould to which it owes its form. The impressions are in gray sandstone.

On the reversed surface of the slab is seen a small piece of broken quartz, about half an inch square. This piece forms a beautiful illustration of a part of the process by which the sandstone rocks are formed.

The second species of the same genus is the Brontozoum Sillimanium. Of this we have three specimens; the tracks have the same general character with the preceding, but are smaller.

The third species of this genus is styled the Brontozoum Loxonyx, from λοξὸς, a bow, and ὄνυξ, a nail,—a curved nail. It is smaller than the Sillimanium, and has the nail set to one side.

The fourth species, still smaller, is the Brontozoum Gracillimum. On this slab the impressions are in relief; viz.: 1st, of Brontozoum Gracillimum; 2d, of Brontozoum Parallelum; 3d, of the track of a tortoise, fourteen inches long, and two wide. Other extensive eminences and depressions, with rain-drops, may be observed on the same surface.

The fifth species is called Brontozoum Parallelum, from the tracks being on a line with each other. Of this there are two specimens, one of them, however, being a single track. On the surface of the other slab there are at least five distinct tracks, one of them being a small new and undescribed species,—thus making the whole number of species of Brontozoum which we possess to be at least six.

The Second Genus of Struthiones is called Æthyopus, from αἴθυια, a gull, and ποὺς, a foot,—gull-footed. This genus is smaller than the Brontozoum Giganteum; and we have two species, viz. the Æthyopus Lyellianus, which is the larger, and two specimens of Æthyopus Minor. All of these are distinguished from the preceding genus by the winged foot, and in the Lyellianus by the shallowness of the impression. The Æthyopus Minor is not always distinguished by the superficiality of its impression. This is sometimes deep. Therefore this character may not be considered a distinctive one, or the Æthyopus Minor might be referred to another genus. Of the two specimens of this latter species, the first is in depression, tridactylous. The depressions are deep with rain-drops, marks of quadrupeds and zoophytes over the whole surface. The ornithichnic impressions are two in number; one superficial, the other very deep. The reversed surface of this slab contains one tridactylous impression in relief. The second specimen has three depressions; two of which are superficial, and the third is quite deep, displaying, by a depressed surface, the webbed character of the foot.

GROUP SECOND

We shall take, to characterize this group, the Argozoum, from ἀργὴς, swift, winged.

Of this genus there are two species, the larger of which is the Argozoum Disparidigitatum. It is leptodactylous, and remarkable for the length of the middle toe. We have another species, which is smaller than the last named, and in which the toes are nearly of equal length; hence called Argozoum Paridigitatum.

The other genus of this group is the Platypterna, and our specimen is named Deaniana. This genus is remarkable for the width of the heel; hence the name, from πλατὺς, broad, and πτέρνη, a heel. It has three toes like the other genera of this group.

GROUP THIRD

This and the succeeding group are tetradactylous; having one toe behind, three forwards.

The third group is leptodactylous; foot usually small, but sometimes of medium size. Of it we have two specimens, viz.: Ornithopus Gallinaceus, and Ornithopus Gracilis. The former is so called from the resemblance to the domestic fowl: for convenience sake, in this and other instances, we use the whole for a part. It is about three inches in length, and the Ornithopus Gracilis about two.

This latter specimen is particularly interesting. It consists of two parts, which open like the covers of a book. These covers present four impressions: first, the superficial, which is distinct, slender, and beautiful— the heel is broad; second, corresponding with this depression and on the inside, is a figure in relief as distinct as the depression; third, on the inside of the second cover is a depression corresponding with the relief last mentioned; fourth, on the outer side is a second relief corresponding with the second depression, but less distinct than either of the other three, still, however, exhibiting three toes pointing anteriorly, but the hind toe is wanting. The whole of this double slab forms a series of cameos and intaglios, measuring four inches by three, and in thickness an inch and a quarter.

GROUP FOURTH

Of the fourth group we have five specimens. The Triænopus, so called from its resemblance to a trident, has besides three leptodactylous toes pointing forwards, a fourth extending backwards in a remarkable way, like the handle of a trident; the impression, however, being expanded so as to show an extensive displacement of the mud. All the specimens of Triænopus are in a beautiful red shale, very thin and fragile, but presenting well-defined impressions, generally about three inches long.

There are two species to this genus. Of the Triænopus Emmonsianus we notice three impressions in relief. In another specimen there is the appearance of a part of the toes of the Anomœpus Scambus, and on the upper side are seen two excavations corresponding with the three impressions. In the last slab, the track of the Triænopus Baileyanus appears to have been made by two feet placed successively in the same spot, which led President Hitchcock to suspect it might have been made by a quadruped. One of the specimens has the Triænopus tracks intermixed in a peculiar way with other impressions.

The specimen representing the genus Harpedactylus is larger than the preceding; and, though leptodactylous, the toes are much broader and also more curved, whence the name Harpedactylus, sickle-finger, from ἅρπη and δάκτυλος.

GROUP FIFTH

The fifth group differs much from the four previous ones. In this and the following groups we pass from the vestiges of birds to those of other animals, some of which are bipeds, some quadrupeds. Many impressions are without any distinct character, belonging probably to the lower animals, to vegetables, and unorganized bodies.

The fifth group comprehends the tracks of an extraordinary animal, the Otozoum.[C] The name which has been given to it is taken from that of an ancient giant, Otus, who with his brother Ephialtes, according to heathen mythology, made war with the gods. These fabled giants were, at nine years of age, nine cubits in width and nine fathoms in height.

The foot is divided into four toes; the two outer of which seem to be connected by a common basis. The inner toe has three phalanges; the second toe, also three; the third and fourth toes, four each. The first is the shortest, the second longer, the third longest, the fourth shorter than the third. It will appear, then, that this track differs from that of birds in the number of toes pointing forwards; these being four, while in birds the forward toes are only three. There is a difference also in the number and arrangement of the articulations.

The track in our possession is twenty inches long by thirteen and a half inches broad. The rock in which it is imbedded is a dark-colored sandstone. President Hitchcock has a slab showing a regular series of tracks of this animal; the distance between the steps being about three feet, and the tracks equidistant and alternate, which would not be the case if the animal had been quadrupedal. In a quadruped, the horse for example, the hind feet are set down near the fore feet, and sometimes even strike them. Hence it must be inferred that the track in question was that of a biped, or of a quadruped which did not use its fore feet in progression, like a kangaroo. We naturally ask, What kind of biped could this have been? Evidently not a man, the size

27

of the foot being too large to admit such a supposition; nor could it have been a bird, the number of toes and their direction not admitting this hypothesis.

Tetradactylous birds, or those which have four toes, have only three of them directed forwards, and the fourth backwards, generally. There are, however, exceptions; some birds have four toes directed forwards: this is the fact with the Hirundo Cypselus and the Pelicanus Aquilus of Linnæus, or Man-of-war Bird. But the articulations are different in the two animals, birds having regularly two, three, four, and five phalanges, and the spur, where it exists, supported by a single osseous phalanx; whereas the Otozoum has three phalanges in the inner and second toe, four in the third and fourth toes. In this last arrangement, the Otozoum is decidedly different from all known birds. It is not likely to have been a tortoise or a lizard. The kangaroo has four feet, and uses only two in progression, moving forward by leaps; also, like the Otozoum, it has four toes; but the size of the toes does not accord with that of the Otozoum, nor is the structure of the foot the same, so far as we know. It has been suggested by Professor Agassiz, that this animal might have been a two-footed frog. Nature had, in those days, animal forms different from those we are acquainted with; and this might have been the fact with the Otozoum.

GROUP SIXTH

We have in this group a specimen of the track of a four-footed animal, which may have been a frog, though different from ours. The feet are unequal in size, and present a different number of toes. In existing frogs there are four toes in the fore feet, and five in the hind; but, in the specimen before us, the front toes are five in number, and the back toes three. It is called, therefore, Anomœpus, unequal-footed. These impressions are in the red shale of Hadley, and very distinct. In some of them the lower leg is indicated, forming an impression six or seven inches long. The feet being smaller than the legs, the impression made by the latter is more expanded, superficial, and broader, yet still very definite. The opinion of President Hitchcock and Dr. Deane is, that the different impressions of five and three toes are those of the anterior and posterior extremities of one animal, which, from the size of the limbs, might be a frog three feet high.

On the same schist with these footmarks, are other curious impressions. The back of the slab is almost covered with the imprints of rain-drops. In the midst of these is a tridactylous impression, probably of a quadruped, crossed at its root by a single depression, nearly an inch broad, and two and a half long: this seems to form part of another broad superficial impression of about seven by four inches, which is probably also quadrupedal. Other parts present the impressions of nails and worm-tracks. At the opposite end is a deep, smooth, regular excavation, which might have been made by a Medusa.

GROUP SEVENTH

The seventh group contains the impressions of the feet of Saurians or lizards. We have a specimen of quadrupedal marks, with five toes to each foot, about an inch long, which may have been made by these animals. The impressions are small, but very distinct. There are lizards of the present day with five toes, about the size of these impressions; and these may, therefore, be set down as belonging to this order of reptiles. Like a number of the last-named specimens, they are in red shale.

GROUP EIGHTH

The eighth group is assigned by President Hitchcock to the Chelonian or turtle tribe. The slab bearing impressions of Brontozoum Gracillimum has a mark about fourteen inches long and two wide, which may be attributed to the plastron or breast-plate of the tortoise. On the slab from Turner's Falls there is a longitudinal furrow, which might have been made by the tail of a turtle; and in various of our slabs are impressions which we think belong to this tribe. We shall have occasion to notice hereafter remarkable tracks of these animals in the old red of Morayshire, in Scotland.

The most distinct of the traces of chelonians are on the large slab lately obtained for me by President Hitchcock from Greenfield. (Vide Plate.) This interesting slab contains the traces of quadrupeds, various birds, and two trails of chelonians: the largest of these is nearly five feet long, and four inches in diameter. The trail is composed of a number of parallel elevations, comparatively superficial.

GROUP NINTH

Of the ninth group, containing the marks of Annelidæ, Crustacea, and Zoophytes, we have various specimens.

The impressions of insects do not seem as yet to have been distinguished on the ancient rocks. There is reason to believe, however, that many of the marks we discover in the rocky beds might have been made by the feet and bodies of large insects; and small species of the same tribes have been found imbedded in, and actually constituting, immense masses of calcareous and siliceous rocks.

The tracks of worms are numerous. No doubt these worms drew together a concourse of birds to the shores on which they rolled. On various slabs we find long cylindrical furrows, about the eighth of an inch in diameter, and of different lengths; one of them, in the slab from Dr. Deane, being eight or nine inches long. To these impressions the name of Herpystezoum, from ἑρπυστὴς, crawling, has been given. They vary, however, and some of them are very likely to be the tracks of the common earth-worm, or of some species of worm which existed when these rocks were formed. These impressions vary in length and in diameter; some of them are moderately regular, and others irregularly curved.

Very interesting tracks have been found in the ancient Potsdam white sandstone of Beauharnais, on the St. Lawrence, by Mr. Logan, an excellent geologist of Canada, and determined by Professor Owen to belong to Crustacea, crabs. The number of impressions made by each foot is sometimes seven, sometimes eight, and even more. This track, showing the traces of Crustacea, goes to form another link in the chain of fossil footsteps.

The Medusæ, commonly called jelly-fish, dissolving as they do under the influence of the sun and air, would hardly be expected to leave their traces impressed on ancient rocks. Professor D'Orbigny, however, has watched the dissolution of these animals on the sea-shore, and found that, after wasting, they may leave their impressions on the sand; which, not being disturbed by a high tide for nearly a month, retains the impression of the zoophyte, and serves as a mould to receive materials which take a cast and transmit it to subsequent ages. We find one of these impressions on the slab of the Anomœpus Scambus; and President Hitchcock, having examined it, is of opinion that it retains the traces of a Medusa. The impression is about five inches in diameter, of a darker color and smoother texture than the rest of the rock. Its edges fade away gradually in the surface of the subjacent sandstone. A similar impression is found on the superior surface of the slab containing the Argozoum.

GROUP TENTH

The tenth group contains the Harpagopus, a name derived from ἁρπαγὴ, seizure, rapine. It is represented by President Hitchcock as having the form of a drag. The figure given by him resembles in a degree the foot of the African ostrich; being a long thick toe, with a shorter one, not unlike a thumb, on the side. An impression approximating this, but of small size, may be seen on the slab of the Anomœpus Scambus.

The formation of bird-tracks is well represented by a clay specimen, about an inch thick, and ten inches long. This is a piece of dried clay, obtained by President Hitchcock from the banks of the Connecticut, and produced by washings from clay on the shore above, covered with foot-impressions of a small tridactylous bird, and dried in the sun. This piece shows, in a way not to be questioned, the manner in which the ancient vestiges were produced. Sir Charles Lyell noticed a similar fact on the banks of the Bay of Fundy.

ORGANIC IMPRESSIONS

The second great division of fossil impressions is called Organic, meaning impressions made by organized bodies; the bones of animals, fishes, and vegetables.

Near one extremity of the slab of the Ornithopus Gallinaceus is an elevation, about a foot long, and between one and two inches wide, projecting from the surface nearly half an inch. It has the appearance of a round bar of iron imbedded in the rock, which is clayey sandstone. This apparent bar of iron was probably a bone, buried in the stone, now silicified and impregnated with iron; the animal matter having entirely disappeared. In the slab of the Brontozoum Sillimanium is a projection about seven or eight inches long, and half an inch wide; probably the bone of an animal, perhaps a clavicle of the Brontozoum Giganteum.

The vestiges of fishes are very numerous in the sandstone rocks of Connecticut River. We have not less than two dozen specimens from this locality; a number equal to all the other specimens in our collection. These impressions of fishes are generally from three to six inches long, and three or four inches wide. They are of the grand division denominated by Professor Agassiz "heterocercal," having their tails unequally bilobed, from the partial prolongation of the dorsal spine; and they are considered to be of lower antiquity than the fishes which are entirely heterocercal. The most remarkable of the fish-specimens in our collection is a Cephalaspis (?): this fish is found in the specimen containing tracks of the Brontozoum Gracillimum, and traces of a turtle or tortoise. This fossil was discovered in the upper layer of the old red sandstone of Scotland, and had been mistaken by some for a trilobite: to us it appeared to be a Limulus, but further observation leads us to believe it to be a Cephalaspis. It exhibits a convex disc, four inches across, by two inches from above downwards, and a tail at right angles with the disc, the uncovered part of which is three inches long. The animal has been described by Professor Agassiz as being composed of a strong buckler, with a pointed horn at either termination of the crescent, and an angular tail.

To the vegetable impressions discovered among the sandstone rocks a peculiar name has not yet been assigned. When, however, we consider the strong probability that many impressions of stalks, leaves, fruits, and other parts of vegetables, may be hereafter discovered in these rocks, it will be found convenient to have a distinctive denomination. Vast numbers of vegetable impressions of a distinct and beautiful appearance, and in great variety, have been found in the coal-formation, which is nearly allied to the sandstone: such are the Sigillaria, Stigmaria, Equisetaceæ, Lycopodiaceæ, Coniferæ, Cycadeæ, &c. It is sufficient to say that the number of these has been already swelled to many hundreds: we must also believe, that some of the impressions in sandstone rocks which have been assigned to other substances ought to be attributed to vegetables. We may, therefore, venture to call the vegetable impressions "phytological."

A number of our slabs bear impressions of vegetables; either twigs of trees, or spires of plants. In a fragment broken from one of the toes of the Brontozoum Giganteum, we see a cylindrical depression, three inches long, and half an inch in diameter, marked by transverse lines, about the sixth of an inch apart, and presenting an unquestionable appearance of a fragment of a twig of an ancient vegetable, which had been trodden under the foot of the mighty Brontozoum. On the reversed surface of the same slab are found impressions, which were produced by a number of fragments of sticks, five or six inches long, lying at right angles, or nearly so. One of these sticks has been broken, and its pieces are slightly displaced from each other. Various other specimens contain the marks of sticks, or twigs of trees. The striæ, so distinctly discernable in a number of these portions, having been compared with twigs of the existing coniferæ (?), were found to resemble them. Some of these sticks show the appearance of incipient carbonization; yet the rock is sandstone, presenting, as already mentioned, distinct appearances of quartz, and other substances of which the arenaceous rocks are composed.

PHYSICAL IMPRESSIONS

The third great division of impressions in the sandstone rocks is called Physical, meaning those made by inanimate and unorganized substances; such are rain-drops, ripple-marks, and coprolites.

1. Marks of rain-drops, described on page 20, appear to be quite common. We have two or three specimens in relief, and as many in depression. They occur as follows: 1st, on the upper surface of the slab first described; 2d, on that of the Platypterna; 3d, on that of the Æthyopus Lyellianus; 4th, on that of the Brontozoum Gracillimum; 5th, on that of the Æthyopus Minor; 6th, on that of the Anomœpus Scambus; 7th, on the recent clay; also in one small hand-specimen, and in a second containing two fishes. They show that, in those ancient periods when the Brontozoum Giganteum and the Otozoum resided in these parts, showers were frequent, and probably abundant for the supply of the wants and the gratification of the appetites of these animals, then common, but which now appear to us so extraordinary.

2. Ripple-marks are seen in a number of these pieces; for example, on the slab first described, on the Brontozoum Sillimanium slab, on the Brontozoum Gracillimum slab, on one of the Triænopus, and on the upper surface of the Greenfield slab. These marks are represented by parallel curves, or straight lines, distant from each other from half an inch to an inch, and presenting a slight degree of prominence. There is another form of ripple-marks(?), differing from those above described. These are of a circular and mammillary form: they are strewed thickly, like little islets, approximating to each other. They are seen distinctly on one of the slabs of the Brontozoum Sillimanium, on that of the Æthyopus Lyellianus, and some others. Whether they are to be considered as accumulations of sand and clay, formed by the action of the sea, we are uncertain; but there seems to be no other cause to which they can be assigned with so great

probability.

3. Coprolites, the fossilized ejections of animals, are intermixed with other animal vestiges in the sandstone of Connecticut River, and afford additional proof of the former existence of animals about these rocks.

The latest accounts of fossil footprints we have had occasion to notice are those of the Crustacea, already mentioned, as found in Canada, and of the Chelonian in Scotland. The Canadian impressions, called by Professor Owen Protichnites, were discovered in the year 1847, and were laid before the London Geological Society in 1851. The most remarkable circumstance about them was their existence, as already stated, in a white sandstone, near the banks of the River St. Lawrence, at Beauharnais. This sandstone, which has been described by New York geologists under the name of Potsdam, is thought to belong to the Silurian system, and to have a higher antiquity than even the "old red."

The Scotch footsteps are situated in the old red sandstone, and are those of a Chelonian. So that we have now two series of tracks, the Crustacea in Canada and the Chelonian in Scotland, of higher antiquity than any which had been previously discovered.

On a review of the labors of President Hitchcock, we are struck with admiration at the immense details that, in the midst of arduous official and literary duties, he has been able to go through with in the period since the foot-tracks were discovered on Connecticut River. Although his labors should be modified by succeeding observers, Science must be ever grateful to him for laying the foundation, and doing so much for the completion, of a work so great, novel, and interesting.

This inquiry seems to us to promise a rich variety; and we hope that President Hitchcock and other observers will continue to explore and cultivate it with undiminished zeal.

DESCRIPTION OF THE PLATE

We are indebted to Photography for enabling us to represent the remarkable slab from Greenfield, and its numerous objects, in a small space, yet with perfect accuracy. This slab is four feet seven and one-half inches in one direction, and four feet one inch transversely to this; in thickness it measures about an inch. It is composed of gray sandstone, in which the micaceous element is conspicuous, and contains many interesting impressions on both surfaces.

The most interesting surface is the inferior; and the objects are, of course, presented in relief. They are, first, two Chelonian tracks; second, four sets of bird-tracks; third, footsteps of an unknown animal. The Chelonian tracks are two in number: the longest measures four feet ten inches; the shorter, two feet nine inches. Both of these impressions are made apparently by the plastron of the turtle. They are from four to eight inches in width, and composed of elevated striæ. These striæ are formed by raised lines, pursuing a course generally regular, but accompanied with some inflections: they are, as the plate represents, very distinct. The shorter track appeared to me to be crossed by another; but the photographic impression, though only a few inches long, enabled me to ascertain that this appearance was produced by bird-tracks above and below.

The bird-tracks are all tridactylous. The first set lies above and to the right of the shorter turtle-track, and is composed of only two steps, proceeding in the course of the plate downwards. The second set of bird-tracks has five impressions, extending from the right superior pointed angle of the slab across the small turtle-track to the larger, in which it is lost. The third set of bird-tracks begins by an impression larger than any other on the piece at the left extremity of the longer turtle-track; and the remainder, three in number, descending towards the right, are the least distinct of any. The fourth set of bird-tracks begins below the longer turtle-track, and ascends by four impressions, crossing the track till it meets the first.

The most curious track, consisting of six digitated impressions, still remains. The first is seen on the left of the longer turtle-track, near the largest bird-track; the second is on the track; the third is above the track; the others cross the slab by fainter impressions. Each of them is composed by two feet, and each foot contains four toes, which are seen more distinctly in some impressions than in others. The largest of these double tracks is about three inches in diameter. Perhaps it would be useless to speculate upon what kind of animal they were made by. There is a similarity between these and the tracks of the Anomœpus Scambus, spoken of in the sixth group. In the latter, however, the toes are five and three. Some experienced persons think they are tracks of the mink, Mustela Lutreola, an animal common at the present day in these parts. This has five toes; but it may be in this as in some other digitigrades, that one of the toes in each foot does not make an impression; or perhaps it is safer to believe, till further investigation is made, that it was an animal of a construction not now existing.

The direction of these tracks presents a puzzle we are not able to unravel; it exhibits the impressions of four toes, and we have supposed it might possess five. In either of these cases, we have no right to consider it a bird-track, but probably a reptile or a mammal. Admitting this to be the fact, we are unable to account for the direction of the steps, which is not alternate, as in the quadruped, but in straight lines. In other words, this animal, supposed to have four legs, gives us the impressions of two only, and both of these placed together.

When the tridactylous tracks are attentively considered, compared with each other, and with the digitated tracks, they appear to exhibit the character of the impressions of the feet of birds so very decidedly, that it would require something more than a philosophic incredulity to question their ornithic origin.

The other side of this slab contains interesting impressions. In the first place, this surface is covered with ripple-marks, each about two inches broad, extending with various degrees of distinctness across the slab, and having an interval of an inch. The width of the ridges is greater than in any of the specimens we have seen.

This surface is almost covered by rain-drops. It has also, among other impressions, one which has been drawn by Mr. Silsbee, our photographist, and represented by the figure below of its proper size. This figure, nearly four and a half inches in length, is an exact resemblance in form, but not in size, of the great Otozoum, as depicted by President Hitchcock, and shown by the actual impression, in our hands, of the great foot, twenty inches long, and of proportionate breadth. The form of the heel, or posterior part of the foot, is the same in the two figures; the toes are equal in both, viz. four in number; the two internal toes correspond in their articulations, and the two external are nearly alike, with a little allowance for a different amount of adipose texture. Whether this was the impression of an infant Otozoum, I pretend not to determine: the drawing was taken by a gentleman who knew nothing of the Otozoum. There are similar impressions, smaller than that last described, on the same surface.

The stone, though now very hard and intractable, having resisted all the chemical agents we could employ, must have remained in a soft state for some time; for the impressions of the foot shown below penetrate to the opposite surface.

In this description we have not attempted to point out all the objects worthy of interest on both sides of this curious slab. Every part of it is full of interest, and presents a field for protracted observations. The surface represented in the plate may, by the aid of a magnifier, be studied without the presence of the stone itself; for the photographic art displays the most minute objects without alteration or omission.

FOOTNOTES

[A] The following are the names of French travellers, who have been supposed to have seen the eggs of the Epyornis in the Island of Madagascar: M. Sganzin, in 1831; M. Goudot, in 1833; M. Dumarele, in 1848; and M. Abadie, in 1850.

[B] Mr. Marsh was a mechanic of the town of Greenfield, and procured his subsistence by his daily labor. Being employed by Dr. Deane in obtaining the sandstone slabs of Ornithichnites, he acquired a taste for the pursuit, entered into it with extraordinary ardor, and accumulated by his own labors a great collection of fine specimens. He unfortunately fell into a consumption, and died in 1852. The collection was sold at public auction for a sum between two and three thousand dollars. The specimens were purchased by the Boston Society of Natural History, by Amherst College, and by various colleges and scientific associations in this country.

[C] The specific name of Moodii has been attached to the Otozoum, from its having been discovered by Mr. Moody.

www.ingramcontent.com/pod-product-compliance
Lightning Source LLC
Chambersburg PA
CBHW071009180526
45168CB00003B/1356